MUSIC FROM ANOTHER SPECIES

MUSIC FROM ANOTHER SPECIES

Australian Birdsong Transcriptions

Hollis Taylor

INDIANA UNIVERSITY PRESS

This book is a publication of

INDIANA UNIVERSITY PRESS
Herman B Wells Library
1320 East 10th Street
Bloomington, Indiana 47405 USA

iupress.org

For customers in the European Union with safety or GPSR concerns, please contact Mare Nostrum Group B.V., Mauritskade 21D, 1091 GC Amsterdam, The Netherlands. Email: gpsr@mare-nostrum.co.uk

Manufactured in Canada

First Printing 2026

LIBRARY OF CONGRESS CATALOGING-IN-PUBLICATION DATA

Names: Taylor, Hollis, author.
Title: Music from another species : Australian birdsong transcriptions / Hollis Taylor.
Description: Bloomington : Indiana University Press, 2026. | Includes bibliographical references and index.
Identifiers: LCCN 2025054913 (print) | LCCN 2025054914 (ebook) | ISBN 9780253075840 (hardback) | ISBN 9780253075864 (epub)
Subjects: LCSH: Birdsongs—Australia. | Butcherbirds—Australia. | BISAC: MUSIC / General | MUSIC / Ethnomusicology
Classification: LCC QL698.5 .T 2026 (print) | LCC QL698.5 (ebook)
LC record available at https://lccn.loc.gov/2025054913
LC ebook record available at https://lccn.loc.gov/2025054914

Contents

Musical Examples *vii*

1. Transcription 1
2. Side by Side: Solo Songs in the Red Centre 6
3. Side by Side: The Neighbors Up North 37
4. The Art Star at a Single Site Across Time 70
5. Additional Solo Songs at a Single Site Across Time 121
6. Elegance and Complexity from Subspecies *Cracticus nigrogularis* 168
7. Elegance and Complexity from Subspecies *Cracticus nigrogularis picatus* 203
8. Duets and Larger Ensembles 245

Acknowledgments *285*

Appendix 1: Audio Tracks *287*

Appendix 2: Notation Conventions *289*

Bibliography *291*

Index *293*

Musical Examples

2.1 Trephina Ridges Music Camp 0.0, 2016: Mount Benstead Creek 9
2.2 Trephina Ridges Music Camp 1.0, 2016 11
2.3 Trephina Ridges Music Camp 2.8, 2016 13
2.4 Trephina Ridges Music Camp 3.1, 2016 14
2.5 Trephina Ridges Music Camp 3.7, 2016 15
2.6 Trephina Ridges Music Camp 4.2, 2016 16
2.7 Trephina Ridges Music Camp 5.3 + 1.3 N on Trephina Gorge Road, 2016 19
2.8 Trephina Ridges Music Camp 5.3 + 5.3 N on Trephina Gorge Road, 2016 21
2.9 Trephina Ridges Music Camp 6.6, 2016 25
2.10 Trephina Ridges Music Camp 7.5, 2016 26
2.11 Trephina Ridges Music Camp 8.0, 2016 28
2.12 Trephina Ridges Music Camp 9.6, 2016: Trephina Creek 30
2.13 Trephina Ridges Music Camp 11.0, 2016: Bitter Springs Creek 31
2.14 Trephina Ridges Music Camp 15.7, 2016 34
2.15 Trephina Ridges Music Camp 18.4, 2016: Ross River Campground 36
3.1 McDonald Street @ Bridge Street, Charters Towers, 2014 37
3.2 Apex Park, Charters Towers, 2014 39
3.3 Back Road, Charters Towers, 2014 40
3.4 Black Jack Road, Charters Towers, 2014 42
3.5 Centenary Park, Charters Towers, 2014 44
3.6 Chloride Street @ Millett Street, Charters Towers, 2014 45
3.7 Gregory Developmental Road @ Hicks Street, Charters Towers, 2014 48
3.8 Hewett Street, Charters Towers, 2014 50
3.9 Louisa Road @ Black Jack Road, Charters Towers, 2014 53

3.10 Louisa Road #2, Charters Towers, 2014 56
3.11 Manners Street @ Hicks Street, Charters Towers, 2014 58
3.12 McPherson Street, Charters Towers, 2014 61
3.13 Prior Street, Charters Towers, 2014 63
3.14 Rutherford Lane @ Towers Street, Charters Towers, 2014 64
3.15 Sandy Creek Road, Charters Towers, 2014 66
3.16 Wellington Road, Charters Towers, 2014 68
4.1 Araluen Arts Centre, 2007 74
4.2 Araluen Arts Centre, 2010 76
4.3 Araluen Arts Centre, 2012 79
4.4 Araluen Arts Centre, 2013 95
4.5 Araluen Arts Centre, 2014 102
4.6 Araluen Arts Centre, 2015 104
4.7 Araluen Arts Centre, 2016 105
4.8 Araluen Arts Centre, 2017 110
4.9 Araluen Arts Centre, 2018 112
4.10 Araluen Arts Centre, 2019 113
4.11 Araluen Arts Centre, 2020 115
4.12 Araluen Arts Centre, 2021 117
4.13 Araluen Arts Centre, 2022 119
4.14 Araluen Arts Centre, 2023 120
5.1 Newhaven Wildlife Sanctuary, 2010 123
5.2 Erija Street, Alice Springs, 2013 128
5.3 Erija Street, Alice Springs, 2014 129
5.4 Erija Street, Alice Springs, 2015 131
5.5 Erija Street, Alice Springs, 2016 133
5.6 Erija Street, Alice Springs, 2017 136
5.7 Erija Street, Alice Springs, 2018 137
5.8 Erija Street, Alice Springs, 2019 139
5.9 Erija Street, Alice Springs, 2020 141
5.10 Erija Street, Alice Springs, 2021 142
5.11 Erija Street, Alice Springs, 2022 143
5.12 Erija Street, Alice Springs, 2023 144
5.13 Greens Park, Georgetown, North Queensland, 2012 145
5.14 Greens Park, Georgetown, North Queensland, 2013 147

5.15 Greens Park, Georgetown, North Queensland, 2014 149
5.16 Greens Park, Georgetown, North Queensland, 2015 152
5.17 Greens Park, Georgetown, North Queensland, 2016 156
5.18 Greens Park, Georgetown, North Queensland, 2017 159
5.19 Cameron Road, North Queensland, 2010 161
5.20 Cameron Road, North Queensland, 2013 162
5.21 Cameron Road, North Queensland, 2014 164
6.1 Alldridge Road, Croydon, North Queensland, 2013 172
6.2 Banana, Central Queensland, 2005 175
6.3 Broken Hill, New South Wales, 2006 177
6.4 Burke and Wills Roadhouse, North Queensland, 2013 178
6.5 Burleigh Heads Golf Course, Queensland, 2013 180
6.6 Fegen Road, Brisbane, Queensland, 2010 181
6.7 Goldfields Caravan Park, Georgetown, North Queensland, 2013 183
6.8 Gold Scrub Lane Near Brisbane, Queensland, 2013 187
6.9 Green Lake, Victoria, 1997 189
6.10 Gregory Street, Bowen, Queensland, 2010 190
6.11 Midway Van Park, Georgetown, 2013 192
6.12 Mildura Along the Murray River, Victoria, 2006 194
6.13 Mount Surprise Station, North Queensland, 2015 195
6.14 Nelly Bay, Magnetic Island, Queensland, 2005 199
6.15 Normanton Street, Georgetown, North Queensland, 2015 200
6.16 Wilcannia, New South Wales, 2016 202
7.1 Archery Track off Ilparpa Road, Alice Springs, 2010 205
7.2 Arltunga Turnoff @ Ross Highway, East of Alice Springs, 2022 206
7.3 Bullen Road 2.5, Alice Springs, 2020 213
7.4 Cassia Hill, Simpson's Gap, Central Australia, 2020 217
7.5 Flynn Drive, Alice Springs, 2014 224
7.6 Irvine Street, Alice Springs, 2016 226
7.7 Jay Creek, Central Australia, 2016 229
7.8 Jessie Gap, Alice Springs, 2019 230
7.9 Oonchiunpa Road, Alice Springs, 2021 234
7.10 Petrick Road, Alice Springs, 2014 236
7.11 St. Mary's, Alice Springs, 2020 238
7.12 Wogarno Station, Western Australia, 2008 244

8.1 Bitter Springs Creek, Ross Highway, Central Australia, 2014 248
8.2 Palm Valley, Central Australia, 2014 256
8.3 Pebbly Beach, Station Creek, New South Wales, 1983 261
8.4 Kinchega, New South Wales, 1984 262
8.5 Tibooboora, New South Wales, 2012 262
8.6 Stockyard Creek, North Queensland, 2010 262
8.7 Witches Falls National Park, Tamborine Mountain, Queensland, 1977 263
8.8 Hugh River, Central Australia, 2014 264
8.9 Booth Road, Townsville, North Queensland, 2006 266
8.10 Bribie Island and Sandstone Point, Queensland, 2013 268

MUSIC FROM ANOTHER SPECIES

1

Transcription

A CENTURY AGO, POET and musician Henry Tate commended to Australian composers the artistic potential of their native musical resources, including Aboriginal music and the sounds of the bush. Tate itemized birdsong of great promise, including those of the pied butcherbird (*Cracticus nigrogularis*), who "among his numerous chants, gives us a grave and gentle measure . . . readily combinable with other calls in an artistic ensemble."[1] Here, I follow Tate's lead with over a hundred transcriptions of pied butcherbird vocalizations—some "grave and of gentle measure," yes, but many others marked by exuberance, timbral riches, chromatic flourishes abutting large leaps and sharp switchbacks, and much more.

My initial pied butcherbird transcription failed. The year was 2001, during my first trip to Australia. At Wogarno Station, a remote sheep station (ranch) in the continent's west, I was surrounded by a vocal trio—who knew birds sang in trios? Low and slow enough for the human ear to parse, their crystalline voices immediately enchanted me, along with their arresting combinatorial prowess. I hastily jotted down a fragment of their efforts, but the full accounting eluded me. I suspected something remarkable was transpiring in their vocalizations. Not long afterward, I determined to devote myself to their scarcely explored vocal culture in an attempt to pry open the musicality and invention rife in their songs.

Key to getting a grasp on pied butcherbird song is, for me, notating it. My history of transcription dates to my time as a classical violinist living in San Francisco in 1973. A friend lent me a cassette of jazz violinist Stéphane Grappelli, founder of the Quintette du Hot Club de France with guitarist Django Reinhardt. I set about notating a couple of Grappelli's solos. At the same time, I was studying at the Berkeley Center for World Music with South Indian violinist T. N. Krishnan, a virtuoso of the Carnatic tradition. I notated the pieces he taught me and the improvisational solos he played based on them.

I was only getting started. As I became increasingly interested in how the violin played a role in other cultures, I stumbled up against the fact that there were (at that time, at least) only a trifling number of books to guide me. Oral traditions required me to learn by ear or transcribe them—I was inclined to the latter. I went on to notate Texas-style fiddle tunes and become the Oregon State Fiddle Champion. This prompted invitations to form my own band and to play in others, especially with Emmy and Grammy Award winner Mason Williams. Bluegrass, old time, country, Western swing, Cajun, pop, rock . . . and add to these probes an artist residency in the Dominican Republic, where I delved into merengue and Afro-Cuban music. During another artist residency, this time in Budapest, I researched Eastern European folk music, with a side trip to Morocco. I played jazz in clubs in the US and Europe. Klezmer this, tango that—master of all these genres? Not at all. Instead, the mastery was in starting from square one on a new genre of music.

Despite my diverse notational pursuits, the initial failure to notate pied butcherbird vocalizations at Wogarno Station was bound to happen. Absolute pitch and long experience notating unfamiliar genres of human music do not suffice when notating such finely nuanced birdsong phrases. These vocalizations call for more than just careful listening in the field: repeated listening to a multitude of soloists at the desk is essential.

Only a handful of pied butcherbird recordings was available in 2004, so I traveled throughout the Australian mainland in pursuit of these feathered choristers. First, I had to find (and recognize) them (fig. 1.1); then, for my purposes, they had to be singing. More often, it happened in reverse—I would find them *because* they were singing. Since 2005, I have spent on average four months each year recording pied butcherbirds. My keen interest is in their hours-long nocturnal solo songs, which are delivered in something approaching the spring. However, song is dependent on factors like light, weather, temperature, the availability of resources to feed the young, and who knows what else. The "town" birds have access to more stable resources and are more reliable singers, generally commencing their nocturnal songs earlier in the season and singing for a longer duration (up to seven hours).

When I am transcribing, my titles take in area/street and state/town or other location identification as well as the recording date. (Granted, the downside of referring to birds based on their street name is that they may invoke dead white men of perhaps questionable colonial intentions.) In my fieldwork, I detail the full spectacle—things musical, of course, but also extramusical aspects like GPS, habitat, weather, temperature, moon phase, and time of sunrise. I document if the bird takes a break to catch an insect (prompting a short pause in singing) as well as the goings-on stage left and right, such as if a rival soloist enters. Although the soundscape is relatively subdued at night, another species may interrupt or overlap—for instance, the howl of a dingo (*Canis lupus dingo*), the boobook of an owl (*Ninox boobook*), the spirited silver tones of a willie wagtail (*Rhipidura leucophrys*)—or, lest one romanticize my fieldwork, the barking of a domestic dog (*Canis familiaris*), the vrooming of a generator or an airplane, and the rumbling of automobile traffic, not to mention the disturbance from unexpected and unwanted human (*Homo sapiens*) intruders.

Call and *song* are not interchangeable terms. Calls are simple social sounds that possess syntax and are employed for contacting, mobbing, feeding, flocking, nesting, greeting, and more. Are calls language or, at minimum, language-like? Possibly, but my interest lies in songs, which I hear, notate, and study as music.[2] While calls are assumed to have some innate basis, songs are learned.[3] My study and understanding of these vocalizations benefit from wide readings in the biological (and even the social) sciences as well as in the humanities and fall under the rubric *zoömusicology*, "the study of music in animal culture."[4]

Figure 1.1 A pied butcherbird nicknamed "Flapper" in Maleny, Queensland. Photo by Neil Boucher.

After recording the nocturnal song, I stay on for the dawn chorus, a veritable roll call of diurnal avian species. Pied butcherbirds contribute here as well, usually in ensemble song. They readily embrace duos, trios, and larger ensembles—I have filmed as many as eight birds singing together at a single location.

While listening intently, I notate at the desk and not in the field:

> BANKSIA GREEN CAMPGROUND, Myall Lakes National Park, New South Wales, 18 March 2013, 3:22 am: A seven-note pied butcherbird fanfare awakens me. I know the drill—get up, grab the gear, and head out—but it's absolutely quiet again. I also know what the notes were. I could sit up and write them down, which is my instinct, but I have trained myself to only notate when I have a backup recording so I can double-check my work. I settle for experiencing the event without distraction.[5]

Composers wishing to make use of the sonic constructs of this species will appreciate individual choristers' attention to matters of shape and balance, with regular contrast in successive phrases in the parameters of timbre, tessitura, intervallic size, melodic contour, and rhythm. Along with images from my fieldwork, excerpted audio recordings of some transcriptions from this volume are available online. Labeled (♦ Audio 1) and so on, they can be sourced at https://www.piedbutcherbird.com, where they are organized by chapter. Appendix 1 details the available audio tracks. Those not available online can be sourced from the author on request.

When transcribing, I import the birdsong recording into my music notation program to carefully compare matters of rhythm and pitch with the avian model. Each pied butcherbird phrase is followed by a longer period of nonsinging (of several seconds or more). The phrases need to breathe, not the bird—the interphrase interval is musically meaningful.

The choice of a metronomic marking is subjective, and not all phrases fit straightforwardly into one choice. The time signature is equally subjective and perhaps not relevant to the birds themselves. I usually cleave to 4/4 as a straightforward meter for human musicians. Other analysts or composers might well craft alternate approaches to this standard time signature.

When I feel that the phrases are well represented and starting to repeat, I place a solid bar line with an asterisk (*) above it for visual emphasis. Only new phrases and variants then feature in the transcription (and no transcription presents all phrases and variants, since nocturnal songs may see new iterations of both on subsequent nights—see the Newhaven Wildlife Sanctuary transcription in chap. 5). Unlike the earlier phrases that often extend across bar lines, each of these is fitted into a single bar (no matter the length) followed by a double bar; in the interest of legibility, some phrases begin with rests of less than one beat.

The ear is generous and even flexible when it comes to pitch discrimination.[6] Nonetheless, when I encounter microtones that my ear will not smooth out, I rely on half-sharp signs to approximate the intonation. I have been asked more than once whether pied butcherbirds sing in just intonation, which some consider a "natural" intonation for a "natural" species. One scholar suggested that I take frequency measurements to examine if pied butcherbirds sing in simple integer ratios or something approaching just intonation, which I did. My attempts were inconclusive, largely because measuring a pitch, especially one with portamento, is thorny. Should I measure the most intense (loudest) section of a note, the longest duration of the note, or the average frequency of the note? The process was unreliable across multiple applications, reminding us of what Gregory Bateson cautioned: "No data are truly raw, all are dependent on editing by the human animal or his instruments."[7] I decided not to cook. In the end, my goal remains to produce legible, nonspecialized transcriptions.

Occasional deviations from this overarching approach serve to highlight alternative modes of analysis, and I clarify such instances in each chapter's introductory remarks. Appendix 2 further details notation conventions and abbreviations.

We know the hand of J. S. Bach. We know Jesuit polymath Athanasius Kircher's 1650 representation of a rooster, cuckoo, nightingale, and more.[8] We know the hand of Olivier Messiaen, especially his birdsong transcriptions.[9] Less known but more impressive to me are the birdsong notations of Johannes C. Andersen, Heinz Tiessen, Olavi Sotavalta, Péter Szöke, and David Hindley.[10] Even though my transcriptions are digitally formatted, I imagine my hand is unambiguously identifiable in this volume. Fostered through long experience in multiple genres, my transcription modified and improved—but does that make it unimpeachable?

No transcription is absolute or faultless. When I return to a completed transcription, I may, for instance, imagine how a grace note might have replaced a thirty-second note or vice versa. What pitch does that portamento end on—the one I hear or the one I measure in the sonogram window? Listen again; measure again; listen one more time. In any case, music notation is both meaningful and powerful, a heritage technology that says *music*. Pied butcherbird phrases belong in such a setting.

Notes

1. Tate, *Australian Musical Possibilities*, 20.

2. Potentially productive cross-species comparisons of both language and music are typically obscured or suppressed via claims of human exceptionalism. Space does not allow a full examination of this vast and fraught topic, but for more on the musicality inherent in birdsong, see Taylor, *Is Birdsong Music?* For more on whether birdsong is best studied as language or music, including the outsize roles played by function and the gatekeeping capacities of consciousness and intentionality, see Taylor, "Birdsong: Language or Music?"

3. Taylor, "Evidence for Teaching."

4. Taylor, *Is Birdsong Music?*, 4. For more on the field of zoömusicology, see my extensive multidisciplinary bibliographies in Taylor, "How Musical Are Animals?" as well as on my website, https://www.zoomusicology.com. Among these references, two are foundational: Mâche, *Music, Myth and Nature*; and Martinelli, *How Musical Is a Whale?*

5. Taylor, *Is Birdsong Music?*, 65.

6. See, for instance, Seashore, *Psychology of Music*; Sethares, *Tuning, Timbre, Spectrum, Scale*, 75; and Sundberg, "Perception of Singing," 171–214.

7. Bateson, *Steps to an Ecology of Mind*, xx.

8. Kircher, *Musurgia Universalis*.

9. For more on my investigations into Messiaen's birdsong transcriptions, see Taylor, "Olivier Messiaen's Transcription"; Curtis and Taylor, "Olivier Messiaen and the Albert's Lyrebird"; Taylor, "Whose Bird Is It?; and Taylor, "Messiaen's Australian Birds."

10. Andersen, *Bird-Song and New Zealand Song Birds*; Tiessen, *Musik der Natur*; Sotavalta, "Analysis of the Song Patterns," 2; Sző́ke et al., "Musical Microcosm," 431 and 427; and Hindley, "Music of Birdsong."

2

Side by Side

Solo Songs in the Red Centre

Each pied butcherbird sings differently from the next, and they change their songs annually. Neighbors typically sing phrases disparate from one another, or they may partially share repertoire (and certainly they would know their neighbor's repertoire, even if they choose not to sing it). Shared items range from similar motifs to entire phrases, more often delivered as variants than as exact matches.

A preference for additive process sees motifs added and deleted from a phrase like snap-together beads. Sticking with this analogy, the orange bead (any given motif) may begin or end a phrase or be placed somewhere in the middle of it.[1] Despite their fluctuating order of appearance in a phrase, the orange (or purple, or yellow, or . . .) beads appear regularly—they are evenly spaced. We are reminded yet again how foundational repetition with variation is to music. These birds keep interest engaged by varying phrases, but they keep it from getting too confusing for their avian audience by having familiar motifs inserted into various parts of their phrases.

This chapter places side by side the nocturnal solo songs of fifteen neighbors to the east of Alice Springs (Mparntwe) in Central Australia, also known as the Red Centre. The East MacDonnell Ranges (Yeperenye) flank the entire stretch, with an occasional gap or water hole. All the transcribed songs in this chapter hail from the spring of 2016 and are presented from west to east at the conclusion of this chapter's text. In a map (which can be sourced at https://www.piedbutcherbird.com), white dots mark neighbors down an 18.4-km (11.4-mile) stretch along the Ross Highway about an hour's drive east of town, extending from Mount Benstead Creek to Ross River Resort campground. Sites fluctuate slightly each season, and a bird may have several songposts over the course of a night. Many individuals are within earshot of their neighbors, and, in any case, all are a short flight away from one another.

Each year, I enjoy the sonic treasures of this compact but productive series of field sites. Only twice has a car passed in the night, and no generators intrude until the end at Ross River Resort campground. I refer to this stretch as Trephina Ridges Music Camp:

> *Festival music and music camps.* Do pied butcherbirds, like humans, go to music camps, have song festivals, or conduct music tourism? In the spring of 2014, I recorded at six new sites along the Ross River Highway east of Alice Springs, each a few miles from the previous. Due to the large number of energetic individual singers of varying abilities at each site and no lone singers in the miles in between, each "camp" had a celebratory feel. I spontaneously wrote up these zones of beauty in my field notes as "music camps." For instance, two main soloists held forth with rolling phrases in the massive river red gums at Bitter Springs Music Camp, with a number of understudies also joining in. Several miles away and across the highway, Trephina Ridges Music Camp had birdsong coming from several different clumps of [struggling] trees lining a dry creekbed. Afterward, I wondered if I was simply expressing my enthusiasm for the experience, or if there could possibly be something to it. Some bird species are known to gather in leks for courtship during the breeding season. My speculation here goes beyond this, to wondering if there is a function outside of breeding for these "camps," that of coming together to learn and celebrate singing. At first blush it seems preposterous, given everything I have read about the function of birdsong—but if instead of thinking about it I go back to my sense of it, I come to a different conclusion.[2]

Commonalities among some individuals figure in these notations, but more often, one comes upon variations and novel phrases. Some longer transcriptions shed light on matters of timing, variation, and combinatoriality and demonstrate the lack of predictability on the phrase level.[3]

I head east from Alice Springs on the Ross Highway, a road that sometimes requires drivers to share one lane and urgently pull off onto a gravel slope if another vehicle approaches from the opposite direction. More pertinent to my drive is that it be cautiously slow, the driver remaining at full alert for kangaroo, wallaby, owl, camel, and others with whom I share the nighttime road. At 59.9 km (37 miles), I stop to listen and reset the odometer at my Trephina Ridges Music Camp 0.0, known to others simply as Mount Benstead Creek. The word *creek* is aspirational, since it is almost always dry.

At this, the last wide-open space along the road before it narrows (thus giving me better access to birds), distant river red gums are on offer for soloists to choose from as songposts. I often must pass this site by, with disappointingly far-off soloists in rugged terrain that cannot be traversed at night. This year, I am in luck, and I head out to record a bird near the road (♦ Audio 1). The next obliging vocalist is at Trephina Ridges Music Camp 1.0 km (♦ Audio 2), followed by recordings at 2.8 km, 3.1 km, 3.7 km (♦ Audio 3), and 4.2 km. This is a highly compressed account, since once I locate a singing bird, I will stay with that individual until dawn unless they fly too far afield.

At 5.3 km, I sometimes turn left onto Trephina Gorge Road and head north, again resetting the odometer. My most reliable sites up the track toward Trephina Gorge Nature Park are at 1.3 km and 5.3 km. Back on the Ross Highway, soloists may hold forth at 6.6 km, 7.5 km (♦ Audio 4), 8.0 km, 9.6 km, and 11.0 km (which is Bitter Springs Creek). Here, the road widens somewhat.

Multiple soloists, some close and others more distant, some with a rapid rate of delivery and others more intermittent, are competing—or are they simply joining in? I do not know.

A metal grid spans the highway at 15.7 km, meant to keep cattle, wild horses, and camels on one side or the other. In this transcription at Ross River grid, as I identify it, only unique phrases are notated, since the slow and sporadic rate of delivery renders measuring the interphrase interval not useful for our purposes. The highway ends at 18.4 km, where I arrive at the campground at Ross River Resort. Rather than pull in through the gate, which might awaken campers, I instead follow a rough track down toward the dry riverbed to record a soloist. Public art is popular at campgrounds. One night here, as I pulled out my tripod to set up my gear, I noticed a new macramé wall hanging on a fence. Then, I saw a tongue flicking: a seven-foot python had curled around the fence and artfully positioned its head dead center. I quickly got back in the Land Cruiser. With each subsequent visit to this site, the moment rushes back at me.

Notes

1. Janney et al., "Temporal Regularity."
2. Taylor, *Is Birdsong Music?*, 174.
3. For more on pied butcherbird combinatoriality, see Janney et al., "Temporal Regularity."

Example 2.1 Trephina Ridges Music Camp 0.0, 2016: Mount Benstead Creek

Example 2.1 *continued*

Example 2.2 Trephina Ridges Music Camp 1.0, 2016

Example 2.2 *continued*

Example 2.3 Trephina Ridges Music Camp 2.8, 2016

Example 2.4 Trephina Ridges Music Camp 3.1, 2016

Example 2.5 Trephina Ridges Music Camp 3.7, 2016

Example 2.6 Trephina Ridges Music Camp 4.2, 2016

Example 2.6 *continued*

Example 2.6 *continued*

Example 2.7 Trephina Ridges Music Camp 5.3 + 1.3 N on Trephina Gorge Road, 2016

Example 2.7 *continued*

Example 2.8 Trephina Ridges Music Camp 5.3 + 5.3 N on Trephina Gorge Road, 2016

Example 2.8 *continued*

Example 2.8 *continued*

Example 2.8 *continued*

Example 2.9 Trephina Ridges Music Camp 6.6, 2016

Example 2.10 Trephina Ridges Music Camp 7.5, 2016

Example 2.10 *continued*

Example 2.11 Trephina Ridges Music Camp 8.0, 2016

Example 2.11 *continued*

Example 2.12 Trephina Ridges Music Camp 9.6, 2016: Trephina Creek

Example 2.13 Trephina Ridges Music Camp 11.0, 2016: Bitter Springs Creek

Example 2.13 *continued*

Example 2.13 *continued*

Example 2.14 Trephina Ridges Music Camp 15.7, 2016

Example 2.14 *continued*

Example 2.15 Trephina Ridges Music Camp 18.4, 2016: Ross River Campground

3

Side by Side

The Neighbors Up North

THIS CHAPTER DETAILS the 2014 singing of sixteen neighbors in the regional North Queensland town of Charters Towers (population approximately eight thousand), where mining and the beef industry dominate. Most of the soloists in Charters Towers have their own take on a distinct motif not typically heard in other pied butcherbird territories. An individual on McDonald Street @ Bridge Street delivered this basic version of the motif, which I notate below in three possible ways (ex. 3.1).

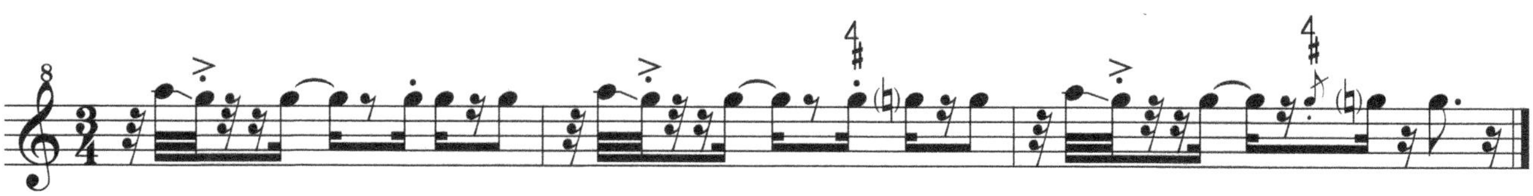

Example 3.1 Three ways to notate a motif commonly heard in Charters Towers, this one from an individual on McDonald Street @ Bridge Street

Variations on this motif all have in common a delivery on G or G# as well as a notable timbral manipulation, which could be described as metallic sounding. On the night of my visit, the McDonald Street @ Bridge Street bird repeated this as an ostinato (♦ Audio 5), offering up no other phrases. We see slight variations of the motif at Apex Park (bar 26), Centenary Park (bar 1), Louisa Road #2 (bar 14), McPherson Street (bars 1–2), Prior Street (bars 47–48), Rutherford Lane (bars 17 and 53), and Wellington Road (bars 2 and 4), with more oblique references from soloists on Chloride Street @ Millett Street (bar 31), Gregory Developmental Road (bars 14 and

33), Hewett Street (bar 1), Louisa Road @ Black Jack Road (notable for variations including bar 1, bars 4–5, and bar 15), Manners Street (the second half of bar 1), and Sandy Creek Road (bars 3–4 and 44). Audio files feature soloists from Chloride Street @ Millett Street (♦ Audio 6), McPherson Street (♦ Audio 7), and Rutherford Lane (♦ Audio 8).

The apparent absence of a tonic or tonal center in many of these birdsongs could be interpreted as a lack of harmonic discernment (a failure to understand the "rules" of Western harmony). Alternatively, it could be understood as a preference for a freer order or, given their evident capacity for absolute pitch, as musical sophistication—they do not require a tonic to anchor their phrases.

Example 3.2 Apex Park, Charters Towers, 2014

Example 3.3 Back Road, Charters Towers, 2014

Example 3.3 *continued*

Example 3.4 Black Jack Road, Charters Towers, 2014

Example 3.4 *continued*

Example 3.5 Centenary Park, Charters Towers, 2014

Example 3.6 Chloride Street @ Millett Street, Charters Towers, 2014

Example 3.6 *continued*

Example 3.6 *continued*

Example 3.7 Gregory Developmental Road @ Hicks Street, Charters Towers, 2014

Example 3.7 *continued*

Example 3.8 Hewett Street, Charters Towers, 2014

Example 3.8 *continued*

Example 3.8 *continued*

Example 3.9 Louisa Road @ Black Jack Road, Charters Towers, 2014

Example 3.9 *continued*

Example 3.9 *continued*

Example 3.10 Louisa Road #2, Charters Towers, 2014

Example 3.10 *continued*

Example 3.11 Manners Street @ Hicks Street, Charters Towers, 2014

Example 3.11 *continued*

Example 3.11 *continued*

Example 3.12 McPherson Street, Charters Towers, 2014

Example 3.12 *continued*

Example 3.13 Prior Street, Charters Towers, 2014

Example 3.14 Rutherford Lane @ Towers Street, Charters Towers, 2014

Example 3.14 *continued*

Example 3.15 Sandy Creek Road, Charters Towers, 2014

Example 3.15 *continued*

Example 3.16 Wellington Road, Charters Towers, 2014

Example 3.16 *continued*

4

The Art Star at a Single Site Across Time

Pied butcherbirds can live up to thirty-five years and are not migratory. However, since I am not working with banded birds, it is unknown whether the individual singing annually at the Araluen Arts Centre (where I first recorded in 2007) is the same vocalist each year. My musician's hunch is that it is indeed the same individual and that 2023 was his final season of singing (I heard no nocturnal singing at this site during my 2024 fieldwork). Songs in this chapter are presented chronologically.

This location is a difficult one to record at, since it is on a major road. Even at night, there can be foot and vehicular traffic:

> Araluen Arts Centre, Alice Springs, 7 October 2012, 3:00 a.m.: This cultural arts precinct is located on the Two Women Dreaming Track, with seven registered Aboriginal sacred sites and trees of significance, including a three-hundred-year-old corkwood tree in the sculpture garden. Pied butcherbirds undoubtedly appreciate this tree, as well as the massive river red gums throughout the property. The soloist that I record here annually is a favorite of mine, but when the bird begins, the recorder picks up about equal measures of bird and humans—a young couple is fighting nearby. I take sides, but only in my mind. The argument is too heated to approach them and request that they move along. I'll just have to make do. Then, as if our species could not do any worse, a man wearing a blanket paces back and forth nearby, giving me a furtive glance from his parallel universe. The couple's argument begins to get physical, and I wonder if I might have to intervene when they head my way. I'm torn between them and my recording and worried that my presence could increase the situation's volatility. Meanwhile, the pacer lies down on the ground next to me, requiring yet another personal safety assessment—all in a night's work.[1]

I am not alone in my challenges here. Recently, my partner and occasional musical collaborator, the violinist/composer Jon Rose, recorded video and audio of the bird I call "Araluen" overnight while I recorded at another site. These feature in a work for the Kronos Quartet commissioned by the Adelaide Festival.[2] The bird kept moving songposts, so Jon found himself far removed from the car when three apparent "good guys" (GG) and one very drunk guy (DG) appeared in the parking lot at 4:00 a.m.:

GG: Good morning.
GG: I'm sorry, brother.
J: Brother . . .
GG: Merry Christmas!
J: I'm just recording.
DG: What you recording?
J: The bird, but he's just gone.
DG: Can I get a cigarette?
J: No, I don't have any cigarettes. I'm sorry.
DG: What you doing out here? It's late at night.
J: Recording the bird.
DG: It's late at night.
J: Recording the bird.
DG: It's late at night.
J: Can you hear him?
DG: Can you give me some money?
J: No, I don't have any money.
DG: Twenty-five—twenty-five dollars.
J: I don't have any money.
DG: Twenty-five dollars.
J: I don't . . .
GG: Listen it! He's trying to, he's trying to record . . .
J: I'm trying to record the bird.
GG: He's trying to record the early bird, early bird.
DG: Yes, yes, yes.
GG: Sorry to . . .
DG: What's the time, what's the time right now?
J: Have a good night.
DG: What's the time right now?
J: Have a good night.
DG: What's the time right now? It's four o'clock in the morning—what the fuck you doing out here?
J: Recording the bird—that's what I do. I study birds.
DG: You'll get robbed out here.
GG: Nah . . .
DG: You'll get robbed out here, brother.
GG: He's alright, he's alright.
DG: You'll get robbed out here.
J: I'm doing something nice.

DG: Don't fuck around, bro. . . . You'll get robbed out here.
GG: It's all cool.
DG: You'll get robbed out here.
GG: It's all cool.
DG: You'll get robbed out here. . . . Motherfucker.
GG: Bye, brother.
DG: You'll get robbed out here, motherfucker.
GG: It's alright. Take care, mister. It's alright, it's alright. He's my cousin. He's just, you know, he's just drunk late at night. He's had, he's had a . . .
DG: You'll get robbed.
GG: He's had a, he's had a . . .
DG: And let me tell you, let me tell you, let me tell you. This whole equipment right here you got.
J: It's for the bird. I study the bird.
DG: People like me, black people like me, they'll rob you for this shit.
GG: Yeah.
J: I know . . .
GG: Try to take care of yourself.
DG: I'm just trying to tell you.
J: Okay, well, thank you.
GG: We're alright, we're good, we're good criminals, too. We're not criminals. We're alright. It's alright, it's alright. Merry Christmas.
J: Thank you.
GG: Sorry about that.
J: No worries.
GG: Have a good morning, too.
J: You too, mate.

Everything about recording here is a test for the recordist, one typically well rewarded, at least. Audio files from 2014 (♦ Audio 9) and 2017 (♦ Audio 10) demonstrate the strong vocalizations of this soloist, who sings not just well but long in the season and begins early in the night (often commencing at 1:30 or 2:00 a.m.). My field notes from 2014 read:

> He was clearly visible; he did not move his head up and down as in ensemble songs. He kept his head stable although he looked around between phrases. A YTM [yellow-throated miner (*Manorina flavigula*)] got on the branch twice as if to irritate him, but the pied butcherbird did not stop or even look at the bird no matter how close the YTM came.

In the vernacular, he's a dude.

When the rest of the town birds have not yet commenced for the season, I will sometimes record this individual nightly for a week or more. Such was the case in 2018, where my field notes on a four-plus-hour song read:

> 3:45 a.m., Araluen Arts Centre parking lot, bottle thrown in my direction on my drive in, already singing, four songposts, combinatorial, timbral including one very steep/swift motif, one-pitch motif, fanfare, three-note rising, descending octave, fast paced, rattle, elegant and delicate but strong—like finely wrought iron. I get inside car at 5:15 a.m. due to 3 young men who were loud and rowdy and about 7 seconds' dash from me, feral cat mobbed by YTMs at 5:23 a.m.

With the threat birds and other species face from feral cats in Central Australia, any effort to thwart them is appreciated, even by a pesky yellow-throated miner. Across the street, another strong singer holds forth on Irvine Street (see chap. 7).

One night in 2019, "Araluen" concludes his song with mimicry, including of a telephone ringtone, a car alarm, a dog, a cat, a horse, and a host of Australian birds: Australian magpie (*Cracticus tibicen*), masked lapwing (*Vanellus miles*), peaceful dove (*Geopelia placida*), crested pigeon (*Ocyphaps lophotes*; both the call and the distinctive mechanical sound of their wings in flight), yellow-throated miner, grey shrike-thrush (*Colluricincla harmonica*), black-faced cuckoo shrike (*Coracina novaehollandiae*), and galah (*Eolophus roseicapilla*)—a pink and gray cockatoo much appreciated by humans for their playfulness. Approximately 10 percent of solo songs are appended with mimicry, a full-on cut-and-paste mishmash that typically lasts between thirty and ninety seconds.

Notes

1. Taylor, *Is Birdsong Music?*, 112.
2. "Adelaide Festival 2023," Adelaide Festival, accessed July 7, 2024, https://2023.adelaidefestival.com.au/events/kronos-five-decades.

Example 4.1 Araluen Arts Centre, 2007

Example 4.1 *continued*

Example 4.2 Araluen Arts Centre, 2010

Example 4.2 *continued*

Example 4.2 *continued*

Example 4.3 Araluen Arts Centre, 2012

Example 4.3 *continued*

Example 4.3 *continued*

Example 4.3 *continued*

Example 4.3 *continued*

Example 4.3 *continued*

Example 4.3 *continued*

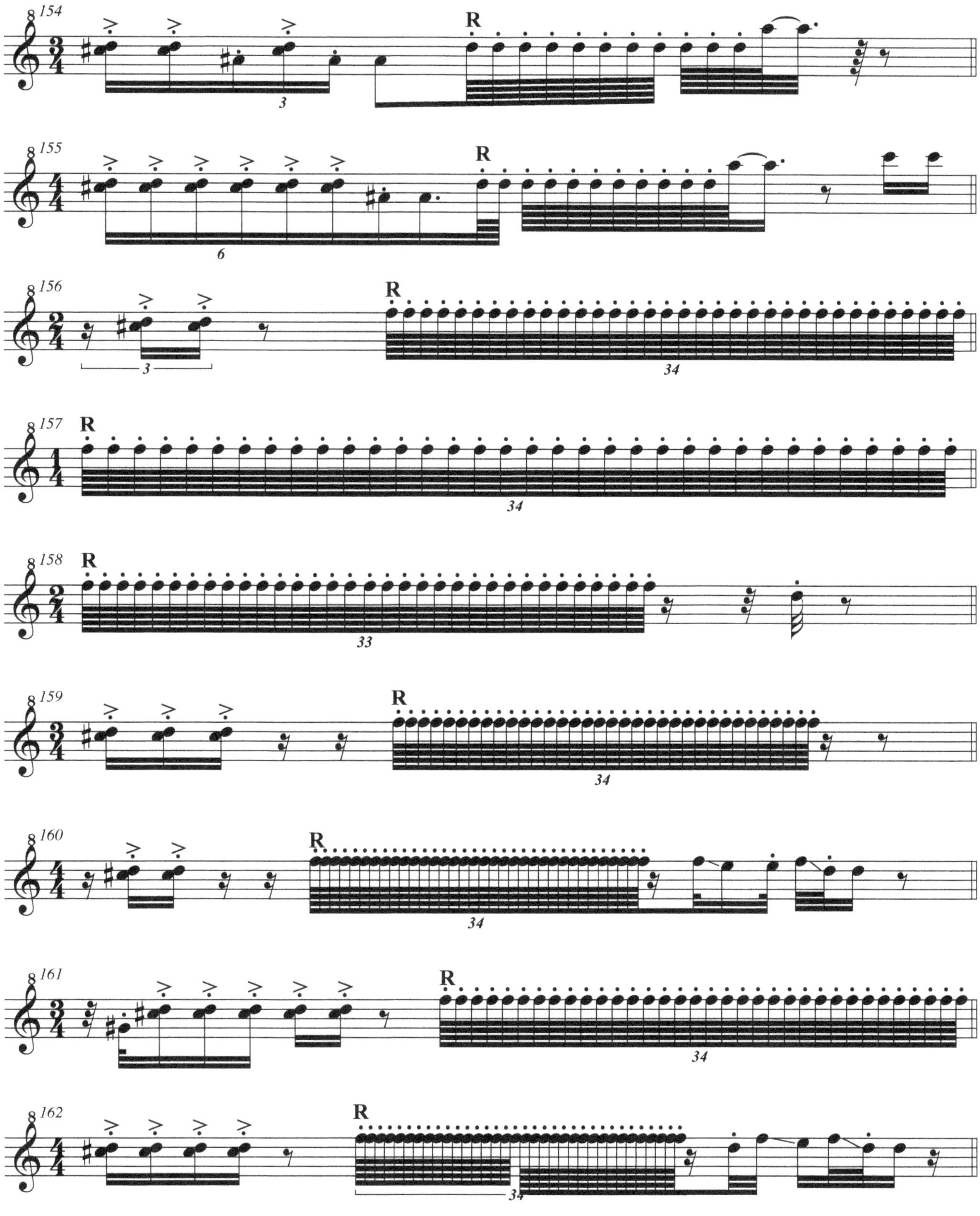

Example 4.3 *continued*

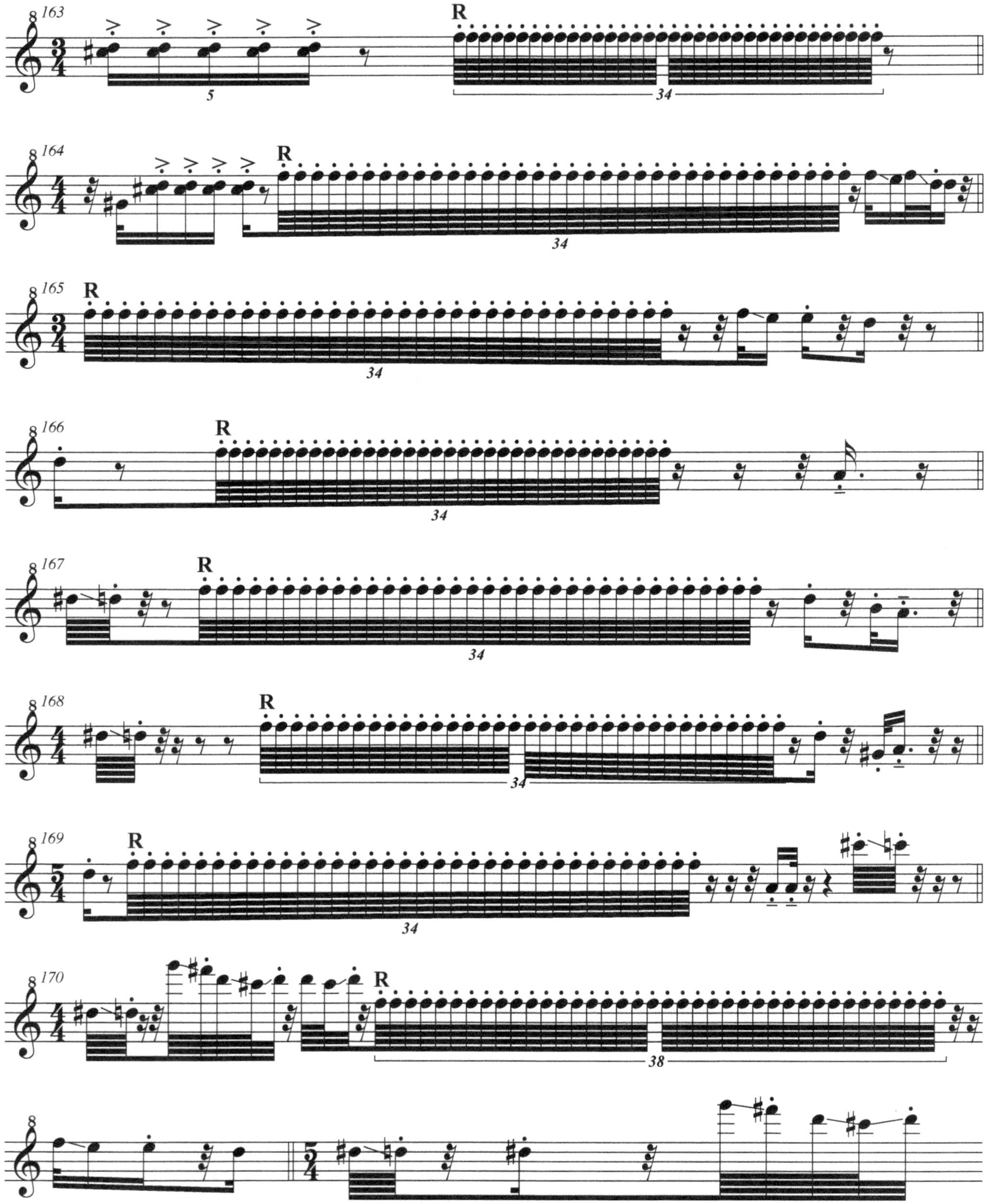

Example 4.3 *continued*

Example 4.3 *continued*

Example 4.3 *continued*

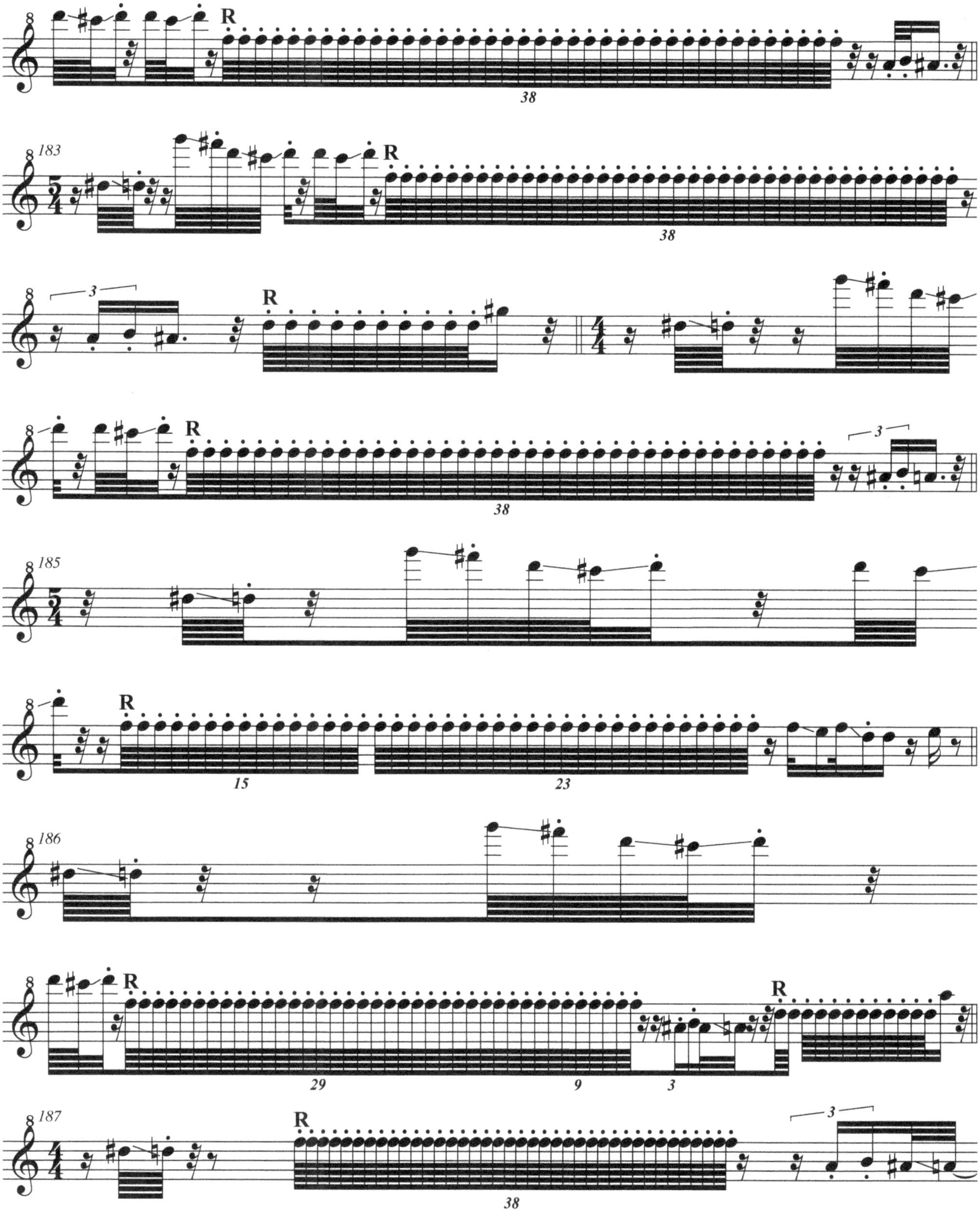

Example 4.3 *continued*

Example 4.3 *continued*

Example 4.3 *continued*

Example 4.3 *continued*

Example 4.3 *continued*

Example 4.4 Araluen Arts Centre, 2013

Example 4.4 *continued*

Example 4.4 *continued*

Example 4.4 *continued*

Example 4.4 *continued*

Example 4.4 *continued*

Example 4.4 *continued*

Example 4.5 Araluen Arts Centre, 2014

Example 4.5 *continued*

Example 4.6 Araluen Arts Centre, 2015

Example 4.7 Araluen Arts Centre, 2016

Example 4.7 *continued*

Example 4.7 *continued*

Example 4.7 *continued*

Example 4.7 *continued*

Example 4.8 Araluen Arts Centre, 2017

Example 4.8 *continued*

Example 4.9 Araluen Arts Centre, 2018

Example 4.10 Araluen Arts Centre, 2019

Example 4.10 *continued*

Example 4.11 Araluen Arts Centre, 2020

Example 4.11 *continued*

Example 4.12 Araluen Arts Centre, 2021

Example 4.12 *continued*

Example 4.13 Araluen Arts Centre, 2022

Example 4.14 Araluen Arts Centre, 2023

5

Additional Solo Songs at a Single Site Across Time

THIS CHAPTER SURVEYS four locations, two from Central Australia and two from far North Queensland. Within each location, the songs are presented chronologically. Although the Queensland birds tend to be quite complex on the page, gesture is the key element in their delivery while exact rhythms are less significant for the human performer—of course, I cannot speak for the avian soloist.

We begin with a five-hour drive from Alice Springs, up the unsealed Tanami Track to the Newhaven Wildlife Sanctuary, where I record a bird in the campground on four successive nights (♦ Audio 11). Even on the final night, he introduces one new phrase type and five variants of other phrases. To explore repertoire size, this transcription shows phrases in order of delivery without interphrase intervals. Since this song submits to such analysis better than most, I also present a typology, or categorization by phrase (A12, D5, E3, etc.).

Erija Street in Alice Springs parallels the Stuart Highway, the major artery in the country running north–south. Even at 3:00 a.m. on a Sunday, when the traffic is diminished, there is still the odd vehicle, plus dogs are carefully listening and eager to bark. I dare not get out of the car lest the dogs start up. The car window is down well in advance, and even though my arm and hand will fatigue from holding the recorder out the window for several hours, that is the only way to work at this site. Recordings here require some filtering, but I find this singer worth the trouble. The transcriptions span eleven successive years, from 2013 to 2023, supported by two audio files, Erija Street 2016 (♦ Audio 12) and Erija Street 2017 (♦ Audio 13). The 2020 phrases, alternating between F and F#, impart a major/minor feeling to the song.

For the next series of transcriptions, we depart Central Australia for Queensland. The Gulf Developmental Road guarantees just a single lane of bitumen. Drivers in this part of far North Queensland must keep constant watch for oncoming traffic and wandering stock and be prepared to urgently pull off onto gravel-and-red-dust shoulders. Acres of termite "tombstones" line much of the road's 442 kilometers (275 miles). I head west for hours, until a long bridge crosses the dry Etheridge River. Here, a cutout of a huge Brahman steer with floppy ears welcomes (and bids farewell to) visitors to Georgetown, which has a population of 243. Grazing has replaced fluctuating mining fortunes, but since the area is in drought, these days grazing is capricious as well. You could say it is an unlikely place to encounter a virtuoso—but Australia is full of surprises:

> Etheridge Riverbed, Greens Park, Georgetown, North Queensland: In 2015, I recorded a bird for five consecutive days. In four of them, this energetic virtuoso of timbral tricks, large leaps, and combinatorics, including two phrases with harmonic implications convincingly paralleling eighteenth-century Western classical music, sang reliably from one songpost. Pull up, point microphone, and wait—it's rarely so simple for me. Day 3 was more typical. I had to go running to stay close as the bird rotated among distant songposts outside the park. Near dawn, the bird flew right over me and landed in a tree ten feet away. I was too close—perhaps for the bird, but definitely for my own sensibility. The signal was so loud that I had to pull the long-distance "shotgun" microphone out of my handheld recorder to avoid distortion. I went to the inbuilt mics. Still too loud. I turned down the input, but what I could not turn down was this virtuoso's high-voltage presence. I took several steps back. The experience was larger than life, much as you might imagine a human soloist, who must project from the stage to the very back of an auditorium, suddenly being right there in your face. It took me back forty-plus years to when I accompanied conductor Walter Susskind to Henryk Szeryng's hotel room to hear a Paganini violin concerto. Szeryng was due to premiere it with the St. Louis Symphony [Paganini's two aged nieces having recently uncovered the score]. His virtuoso technique and the loud, luscious sound he made on his Guarneri del Gesù "Le Duc" were overpowering. At one point, he said while continuing to play, "Paganini himself said this passage was impossible."

Six seasons of singing feature here. The soloist alternates between complexity and elegance. Two audio files support this songster: Greens Park 2014 (♦ Audio 14) and Greens Park 2015 (♦ Audio 15), where the final note of the first phrase features the lowest note that I have recorded by any pied butcherbird. Geckos and fruit bats accompany much of his singing.

The Flinders Highway in North Queensland bears the nickname "Highway of Death," not for the traffic accidents on its roads but for the host of murders and disappearances along it over the past fifty years. I must put this knowledge aside as I prepare to park, alone and in the dark, on a remote stretch of it. Heading west from the coastal city of Townsville (see the duos in chap. 8) but well before Charters Towers (see chap. 3's solo songs), Cameron Road is a short stretch of gravel off the highway before dead-ending at a pastoral property. For several years, I pulled off onto Cameron Road and then struggled to get a fix on a close singer. I finally had to acknowledge that the dominant soloist sings right on the highway, and that was where I would need to park. The shoulder is narrow and steep, the highway is not lit, and stopping at just the right tree in the dark takes careful planning as well as luck. Scant traffic passes in the night, mostly just large trucks; each time they pass, the earth trembles, and the car shakes. Two audio files support this individual (♦ Audio 16 for 2010 and ♦ Audio 17 for 2013).

Example 5.1 Newhaven Wildlife Sanctuary, 2010

Example 5.1 *continued*

Example 5.1 *continued*

Example 5.1 *continued*

Example 5.1 *continued*

Example 5.2 Erija Street, Alice Springs, 2013

Example 5.3 Erija Street, Alice Springs, 2014

Example 5.3 *continued*

Example 5.4 Erija Street, Alice Springs, 2015

Example 5.4 *continued*

Example 5.5 Erija Street, Alice Springs, 2016

Example 5.5 *continued*

Example 5.5 *continued*

Example 5.6 Erija Street, Alice Springs, 2017

Example 5.7 Erija Street, Alice Springs, 2018

Example 5.7 *continued*

Example 5.8 Erija Street, Alice Springs, 2019

Example 5.8 *continued*

Example 5.9 Erija Street, Alice Springs, 2020

Example 5.10 Erija Street, Alice Springs, 2021

Example 5.11 Erija Street, Alice Springs, 2022

Example 5.12 Erija Street, Alice Springs, 2023

Example 5.13 Greens Park, Georgetown, North Queensland, 2012

Example 5.13 *continued*

Example 5.14 Greens Park, Georgetown, North Queensland, 2013

Example 5.14 *continued*

Example 5.15 Greens Park, Georgetown, North Queensland, 2014

Example 5.15 *continued*

Example 5.15 *continued*

Example 5.16 Greens Park, Georgetown, North Queensland, 2015

Example 5.16 *continued*

Example 5.16 *continued*

Example 5.16 *continued*

Example 5.17 Greens Park, Georgetown, North Queensland, 2016

Example 5.17 *continued*

Example 5.17 *continued*

Example 5.18 Greens Park, Georgetown, North Queensland, 2017

Example 5.18 *continued*

Example 5.19 Cameron Road, North Queensland, 2010

Example 5.20 Cameron Road, North Queensland, 2013

Example 5.20 *continued*

Example 5.21 Cameron Road, North Queensland, 2014

Example 5.21 *continued*

Example 5.21 *continued*

Example 5.21 *continued*

6

Elegance and Complexity from Subspecies *Cracticus nigrogularis*

In birdsong, as in fashion, elegance is refusal. Pied butcherbirds have inhabited the planet for some thirteen million years, along the way developing methods for managing a limited palette of pitches and rhythms set in repetitive, combinatorial patterns. Their vocalizations find ample mutualisms in human sonic structures. For instance, in musical games of the late eighteenth (and early nineteenth) century, ready-made motifs were selected by throwing dice, spinning a top, or choosing random numbers. Known as *ars combinatoria*, systematic arrangements of prefabricated materials were well known to Mozart and Haydn. Beginning with only a few simple component pieces, the possibilities for combinations quickly balloon.[1] In *Musurgia Universalis* (1650), Kircher claims that combinatorics enables a novice to attain compositional skill in a short time.[2] And a bird?

All told, my two most productive areas are Central Australia, in and around the town of Alice Springs, and North and Far North Queensland. These areas encapsulate the two subspecies of pied butcherbird, which are visually identical (or nearly) in the field. *Cracticus nigrogularis* (fig. 6.1) inhabits Eastern Australia, while *Cracticus nigrogularis picatus* (fig. 6.2) is found in Central and Western Australia. I defer to the reader/listener to decide whether the songs of the two subspecies betray marked differences.

This chapter surveys *Cracticus nigrogularis* in Queensland, New South Wales, and Victoria. Songs presented here and in the next chapter are in alphabetical order. Below, I set forth reflections on a number of the soloists and their locations. On Alldridge Road in Croydon, North Queensland, I recorded a bird committed to holding forth despite the modest rock festival in a back field still cranking it out at 5:10 a.m. A pet pig roamed freely in the caravan park nearby. Queensland accounts for over 90 percent of Australia's banana production, but Banana, Queensland, is named instead for a long-lived and hardworking bullock whose moniker referred

Figure 6.1 A pied butcherbird (*Cracticus nigrogularis*) in Burleigh Heads, Queensland. Photo by author.

to the color of his hide.[3] The soloist here proposes a snappy song. The Broken Hill phrases are straightforward. I admire them for their sense of shape and balance.

The Burke and Wills Roadhouse is extremely remote:

> Acres of termite "tombstones" line the narrow road on the drive to Burke and Wills Roadhouse. I pass ten pelicans sharing a waterhole with as many cattle (a completely unlikely assemblage in my imagination), and later I startle half a dozen feral pigs at Firey Creek. A whistling kite (*Haliastur sphenurus*) flies off with four feet of red and white ribbon ripped from the gut of kangaroo roadkill. The long stretch between facilities means most folks make at least a short stop at the Burke and Wills crossroads to fuel up. On arrival, I sit outside on the covered deck with an order of chips and salad. An apostlebird (*Struthidea cinerea*) with a lame leg flies up to beg. These highly sociable birds tend to come in groups of a dozen, and thus the name. I shouldn't, but considering the leg, I take pity and throw a tidbit the bird's way. Immediately, over fifty apostles arrive. What happened to the cliché of a dozen? I've been set up!

Figure 6.2 A pied butcherbird (*Cracticus nigrogularis picatus*) at Wogarno Station, Western Australia. Photo by Chris Tate.

Harsh "tearing sandpaper" voices demand that I share. People are looking at me. Like everyone of guilty conscience, I do the only thing one can do: I look away, pretending I didn't do it. A helicopter lands, a welcome distraction. A young man climbs out and walks to the bar. Three minutes later, he exits with a case of cold beer under each arm and a sagging bag with six bottles of spirits hooked onto his left elbow. He climbs back in and flies off.

The sign on the amenities block reads: "Please keep Ladies Toilet Door closed at all times as brown snakes like to visit very often!!" I secure the door and go to bed, much earlier than anyone else. It's difficult to sleep with so many people playing pool and the jukebox roaring. They sing along on the choruses or, rather, shout. The refrain to the current song is "Whiskey for my men, beer for my horses!" I don't hear a thing on the jukebox that can touch what I'll hear at 4:00 am.

Of course, the toilet door is wide open when I get up in the dark, signaling that women also enjoy their whiskey. There's just a slice of moon. I head to the stockyard across the road, where the soloist sang last season, turn on my recorder, and hope.

> Wills does not disappoint, singing forty-five minutes of thrillingly combinatorial phrases, including truncation, virtuosic leaps and portamentos, rattles, low-pitched notes, timbral tricks, and arpeggiated minor chords. This bird's got style. At times, the longer phrases sound crafted on the spot. I'm right underneath Wills, who stays loyal to one songpost for the entire song. I try not to move and sometimes forget to breathe. Two other pied butcherbirds are singing in the distance. Near the end, the phrases become much softer, and Wills caps off the performance with a divertissement: it's mimicry, with a new sound about every second, including a car alarm. Then, Wills flies across the road and sings a species call. I don't hear another pied butcherbird for the rest of the morning.[4]

The next audio file hails from Fegen Street in Brisbane as it begins to spit rain (♦ Audio 18)—despite the traffic, I never ignore the town birds. Farther north, the Georgetown birds form a small club and can all hear one another: Goldfields Caravan Park (♦ Audio 19)—the territory next to the Greens Park soloist (featured in chap. 5); Midway Van Park (♦ Audio 20)—another bird close to Greens Park; and Normanton Street (♦ Audio 21)—which is close to Midway Van Park. Like Fegen Street, these three recordings require some filtering to remove generators, traffic, and even an insistent Australian magpie.

The coastal town of Bowen (population approximately 10,400) sits in the Central Great Barrier Reef region, near the Whitsundays. The individual on Gregory Street is one of many superlative singers in town. I discovered this musical gold mine when a swollen river overflowed its banks, closing the highway I was traveling up and thus diverting me from my intended destination. At sundown, I was forced to stop in Bowen and sleep in my car at a service station, since the caravan parks close at 6:00 p.m. One night in Bowen became many over the years.

In the state of Victoria, two different recordists shared with me a recording of the Green Lake soloist from 1997 (I believe the Australian Wildlife Sound Recording Group held their annual conference there that year). I notated the song in 12/8 because of the overwhelming triplet feel.[5] Green Lake is the only solo song transcription not based on my own recording.

Returning to North Queensland, the pied butcherbirds on Magnetic Island in the Great Barrier Reef are no more, lost to poison, traps, or some other unknown cause. This individual at Nelly Bay sang for three hours and twenty minutes, and the highly combinatorial song is diurnal, not nocturnal. At one point in the recording, a man approached and sprayed herbicide right around me. The bird was never heard again by me or my friends who live on the island. I conducted an extensive analysis of this song.[6] The transcription runs to ninety pages in length, which the Magnetic Island National Park ranger requested I send to him as a tool to buttress his conservation efforts.

Notes

1. Ratner, "Ars Combinatoria."
2. Knobloch, "*Musurgia Universalis*," 266.
3. "Banana, Queensland," Queensland Government, accessed July 14, 2024, https://www.queensland.com/au/en/places-to-see/destination-information/p-596c087d1886d3db60a48ef3-banana.
4. Taylor, *Is Birdsong* Music?, 183–84.
5. "Australian Wildlife Sound Recording Group," accessed July 8, 2024, https://awsrg.org.au.
6. Taylor, *Is Birdsong* Music?, 129–35.

Example 6.1 Alldridge Road, Croydon, North Queensland, 2013

Example 6.1 *continued*

Example 6.1 *continued*

Example 6.2 Banana, Central Queensland, 2005

Example 6.2 *continued*

Example 6.3 Broken Hill, New South Wales, 2006

Example 6.4 Burke and Wills Roadhouse, North Queensland, 2013

Example 6.4 *continued*

Example 6.5 Burleigh Heads Golf Course, Queensland, 2013

Example 6.6 Fegen Road, Brisbane, Queensland, 2010

Example 6.6 *continued*

Example 6.7 Goldfields Caravan Park, Georgetown, North Queensland, 2013

Example 6.7 *continued*

Example 6.7 *continued*

Example 6.7 *continued*

Example 6.8 Gold Scrub Lane Near Brisbane, Queensland, 2013

Example 6.8 *continued*

Example 6.9 Green Lake, Victoria, 1997

Example 6.10 Gregory Street, Bowen, Queensland, 2010

Example 6.10 *continued*

Example 6.11 Midway Van Park, Georgetown, 2013

Example 6.11 *continued*

Example 6.12 Mildura Along the Murray River, Victoria, 2006

Example 6.13 Mount Surprise Station, North Queensland, 2015

Example 6.13 *continued*

Example 6.13 *continued*

Example 6.13 *continued*

Example 6.14 Nelly Bay, Magnetic Island, Queensland, 2005

Example 6.15 Normanton Street, Georgetown, North Queensland, 2015

Example 6.15 *continued*

Example 6.16 Wilcannia, New South Wales, 2016

7

Elegance and Complexity from Subspecies *Cracticus nigrogularis picatus*

Elegance has its place, but a soloist must demand to be heard—a feat perhaps more easily achieved in the virtuoso lane. On this, however, avian consensus is lacking. Some nocturnal songs favor elegance, others complexity; the phrases of the individual at Greens Park in Georgetown (chap. 5) integrate both. Whatever their character may be, these songs remain notable for their musicality, inventiveness, and beauty.

These next birdsongs from Central and Western Australia continue to enchant me. I chose these dozen finalists from a list of 122, which itself was already exceedingly whittled down. Audio files support songs from the Archery Track off Ilparpa Road in Alice Springs (♦ Audio 22); the turnoff to Arltunga on the Ross Highway, east of Alice Springs (♦ Audio 23); 2.5 km down Bullen Road in Alice Springs (♦ Audio 24); and Oonchiunpa Road in Alice Springs (♦ Audio 25).

I perform the Archery Track birdsong on violin, an instrument that allows for adept reproduction of the portamento in bar two. This song also served as my cell phone ringtone for years. The Jay Creek bird is fond of a two-note, short/long articulation (bars 3/4, 19, 33, 38, etc.), which is frequently delivered (especially, but not exclusively, on F# to F or F to E) in Central Australia and by pied butcherbirds more generally.

Singing at the caravan park across the road from the Araluen Arts Centre (chap. 4) or, more often, a block away on Irvine Street in Alice Springs, "Irvine" seems to have crafted a chimeric motif based on the call of an Australian koel (*Eudynamys cyanocephalus*), a large, long-tailed cuckoo. This parasitic species is quite vocal during the breeding season, insistently repeating (to the great irritation of many humans and probably other species) *koo-el*, which typically ascends a semitone in pitch on each repetition. In this transcription, notice the A to A# ascent in bar 42, which sees a crescendo—a rare occurrence for pied butcherbirds, which makes me believe it is borrowed from a koel.

When I heard the soloist at Jessie Gap near Alice Springs in 2019, I immediately thought *fanfare*. My partner Jon Rose and I turned this into a work for trumpet and trombone.[1] The first phrase of Oonchiunpa Road displays a rare but effective melodic contour from pied butcherbirds: a bold ascent spanning an octave plus a perfect fifth.

I end our solo song investigations where I began, at Wogarno Station in Western Australia, where I first heard an antiphonal trio of pied butcherbirds and wanted to hear and know more. This transcription depicts a sprightly nocturnal solo song. Like other members of this species, the feathered dinosaurs at Wogarno hold traditions not unknown or even marginal in human music practice.[2] Twenty years on, I still want to hear and know more.

Notes

1. "Night Songs," Jon Rose, accessed July 8, 2024, https://www.jonroseweb.com/f_projects_night_songs.php.
2. See Taylor, *Is Birdsong* Music?, 147–79; and Taylor, "How Musical Are Animals?"

Example 7.1 Archery Track off Ilparpa Road, Alice Springs, 2010

Example 7.2 Arltunga Turnoff @ Ross Highway, East of Alice Springs, 2022

Example 7.2 *continued*

Example 7.2 *continued*

Example 7.2 *continued*

Example 7.2 *continued*

Example 7.2 *continued*

Example 7.2 *continued*

Example 7.3 Bullen Road 2.5, Alice Springs, 2020

Example 7.3 *continued*

Example 7.3 *continued*

Example 7.3 *continued*

Example 7.4 Cassia Hill, Simpson's Gap, Central Australia, 2020

Example 7.4 *continued*

Example 7.4 *continued*

Example 7.4 *continued*

Example 7.4 *continued*

Example 7.4 *continued*

Example 7.4 *continued*

Example 7.5 Flynn Drive, Alice Springs, 2014

Example 7.5 *continued*

Example 7.6 Irvine Street, Alice Springs, 2016

Example 7.6 *continued*

Example 7.6 *continued*

Example 7.7 Jay Creek, Central Australia, 2016

Example 7.8 Jessie Gap, Alice Springs, 2019

Example 7.8 *continued*

Example 7.8 *continued*

Example 7.8 *continued*

Example 7.9 Oonchiunpa Road, Alice Springs, 2021

Example 7.9 *continued*

Example 7.10 Petrick Road, Alice Springs, 2014

Example 7.10 *continued*

Example 7.11 St. Mary's, Alice Springs, 2020

Example 7.11 *continued*

Example 7.11 *continued*

Example 7.11 *continued*

Example 7.11 *continued*

Example 7.11 *continued*

Example 7.12 Wogarno Station, Western Australia, 2008

8

Duets and Larger Ensembles

Female song is widespread in songbirds, particularly in places like Australia, where birds reside in their territory year-round rather than migrate. Since pied butcherbird females are indistinguishable in the field from their male counterparts, one can merely speculate on female song. Unison duets are rare, but input from supernumeraries may see the doubling of a motif. Duets can unfurl as remarkably complex, coordinated performances or as apparently loose collaborations of informal (or free) timing:

> In human music, a sequential duet where singers take turns so rapidly and with such precision that they form a single melodic line is known as a hocket, paging back to medieval motets, for example, or antiphonal song (ethologists employ this term for birdsong, as do musicologists in describing the call-and-response alternation of an officiant and a choir). The timing of duets is normally such that without visual contact, the contributions do not parse easily, so it is essential that this be annotated by the recordist in the field. When singing, pied butcherbirds also operate on another communication channel: their multimodal display sees them alternating a standard upright posture with raising the bill high and sinking it on the breast, as well as opening or even flapping wings, all of which assist in part identification. These whole-body motor performances enhance the audio and remind us of music's corporeal and haptic dimensions. Ensemble songs may be repeated, either exactly or with variation, a handful of times before the birds switch to a new one, which may be prefaced by a move to a new tree or other songpost. When the result is not a "tight" hocket but instead a looser ensemble song, a strong sense of collective improvisation may be evoked, like the contrapuntal multi-soloing of Dixieland jazz.[1]

Beyond the analysis and comparison of ensemble vocalizations, I am intrigued by the relationship between ensemble and solo songs, including the reach and reworking of a phrase across territories and years. Collectively held ensemble songs see more stability than individually crafted productions; ensemble songs tend to be pied butcherbird classics—their *musica franca*. My hunch is that the ensemble repertoire feeds the crafting of solo phrases, but there could be motion in the opposite direction as well. No matter what, combinatoriality and contrast guide their musical choices.

Two transcriptions serve to transition us from solo to ensemble song. Both are instances of what biologists term *countersinging*, when two nearby soloists jockey for attention. Countersong may be "matched," when a bird sings a phrase that is similar or identical to what the other just sang, or "unmatched." To match a phrase immediately after a neighbor has sung is considered by biologists to be an act of aggression, and perhaps by musicologists to be unimaginative. Audio supports two bouts of countersinging, one at Bitter Springs Creek (♦ Audio 26) (see chap. 2 for another solo song at Trephina Ridges Music Camp 11.0, Bitter Springs Creek) and the other at Palm Valley (♦ Audio 27).

Short duos from five locations feature next. The ensembles at Pebbly Beach at Station Creek, New South Wales, and Kinchega, New South Wales (♦ Audio 28), were recorded by David Lumsdaine. The first seems to him (and to me) to be a singing lesson, with the student singing just a few notes on the bottom staff while the mature bird proposes melismatic embellishment.[2]

The duo in Tibooboora, New South Wales, sees two octave handoffs (♦ Audio 29), which one notices frequently in pied butcherbird duets, including possibly at Stockyard Creek, although I am unsure how to split the parts (♦ Audio 30). The Witches Falls National Park duet from Tamborine Mountain, Queensland, was recorded by H. Sydney Curtis.

The singing at Hugh River exemplifies antiphonal song, where motifs are contributed leisurely and may or may not receive an answer or prompt further musical commentary. In general, ensemble repertoire may take in ten or more duet types over the course of the day, some of fixed presentation and others seemingly more improvised.

The first two lines of the Booth Road transcription feature a hocket (♦ Audio 31), which I could identify as a duet only if I saw the birds deliver it (which I did). The latter two ensemble songs, with seemingly freer timing, I might describe as antiphons.

The neighboring ensembles from Bribie Island and Sandstone Point, Queensland, get the last word—or, in this case, notes. When the singing is such a tangle that I cannot parse it or even hazard a good guess on the division of labor, I may revert to one staff to at least document the musical outcome. Due to the changing character of the antiphonal contributions, I have not imposed a metronomic marking. As a general guide, the chorale-like series of exchanges beginning in bar three impart a laid-back manner, while the chromatic rattles are full-on attention grabbers that move at speed. The vocalizations are presented in their order of delivery over a two-day period.

The theatrics of nocturnal and ensemble song cannot be fully told in transcription. For instance, in October 2014, the Sandstone Point ensemble withdrew from the project. It happened like this: The nocturnal soloist began at Pebble Beach Commons and led me along several songposts over the course of two hours, ending in a small, wooded park where people walk their dogs. Unusually, I was in luck—I had the place to myself. When the solo ended, I decided to stay for the pied butcherbird antiphons in the dawn chorus. However, it slowly occurred to me that the escalating beak claps and zips from a group of about eight butcherbirds were directed at *me*. This time of year, young birds could be in the nest. All backstage passes have been revoked. I quickly departed through the gate, but the ensemble songs were irresistible, so I snuck back and hid behind a tree. Straightaway, a bird landed on the trunk just above my head and uttered a harsh, piercing "zip" call—an eviction notice. I swiftly complied. Although I have seen pied butcherbirds drive out other species, I was not expecting to be attacked by "my" bird (of course, the opposite pertains—I'm *their* human, celebrating their achievements).

Notes

1. Taylor, *Is Birdsong Music?*, 123–24.
2. Taylor, "Evidence for Teaching."

Example 8.1 Bitter Springs Creek, Ross Highway, Central Australia, 2014

Example 8.1 *continued*

Example 8.1 *continued*

Example 8.1 *continued*

Example 8.1 *continued*

Example 8.1 *continued*

Example 8.1 *continued*

Example 8.1 *continued*

Example 8.2 Palm Valley, Central Australia, 2014

Example 8.2 *continued*

Example 8.2 *continued*

Example 8.2 *continued*

Example 8.2 *continued*

Example 8.3 Pebbly Beach, Station Creek, New South Wales, 1983

Example 8.4 Kinchega, New South Wales, 1984

Example 8.5 Tibooboora, New South Wales, 2012

Example 8.6 Stockyard Creek, North Queensland, 2010

Example 8.7 Witches Falls National Park, Tamborine Mountain, Queensland, 1977

Example 8.8 Hugh River, Central Australia, 2014

Example 8.8 *continued*

Example 8.9 Booth Road, Townsville, North Queensland, 2006

Example 8.9 *continued*

Example 8.10 Bribie Island and Sandstone Point, Queensland, 2013

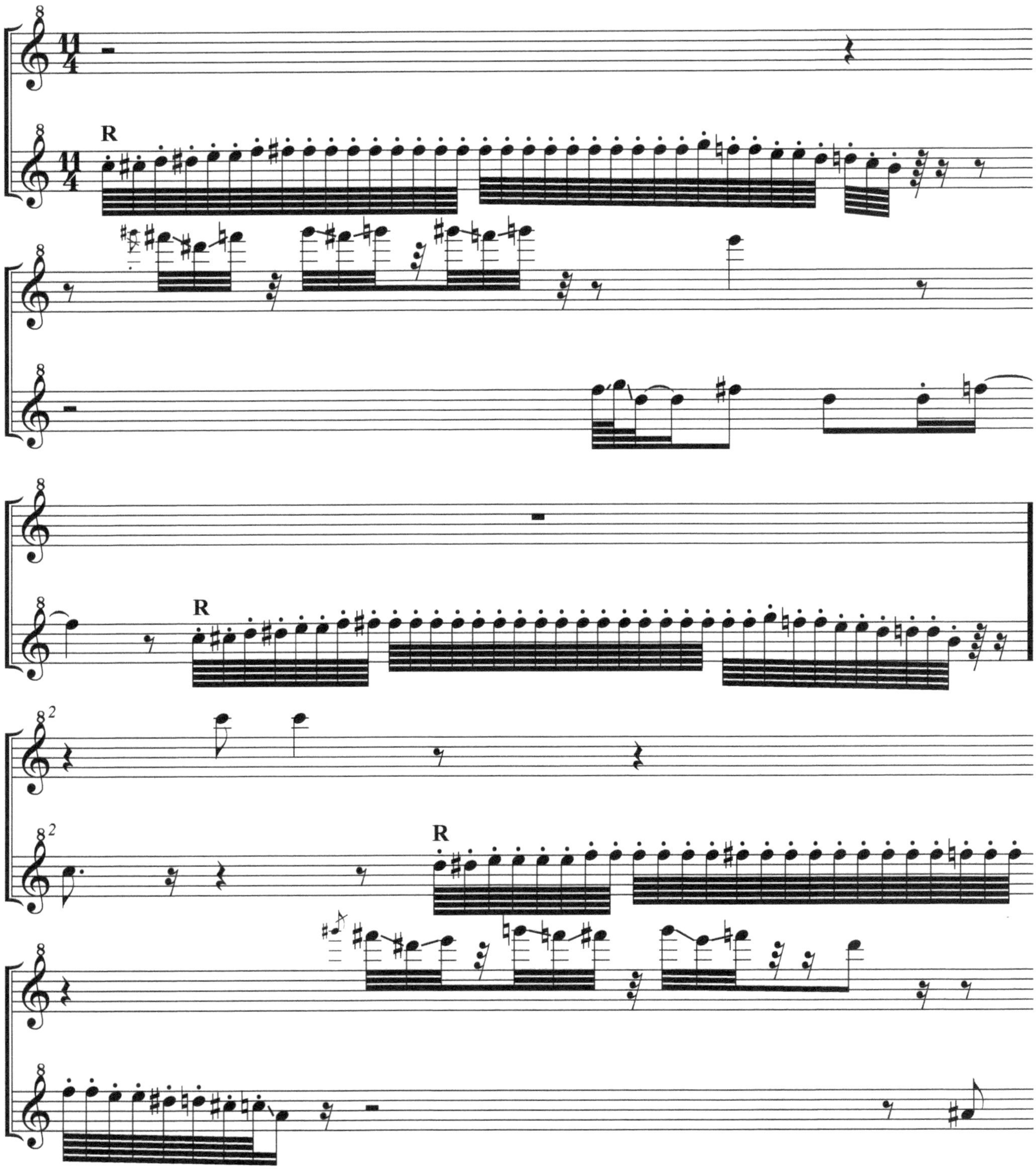

Example 8.10 *continued*

Example 8.10 *continued*

Example 8.10 *continued*

Example 8.10 *continued*

Example 8.10 *continued*

Example 8.10 *continued*

Example 8.10 *continued*

Example 8.10 *continued*

Example 8.10 *continued*

Example 8.10 *continued*

Example 8.10 *continued*

Example 8.10 *continued*

Example 8.10 *continued*

Example 8.10 *continued*

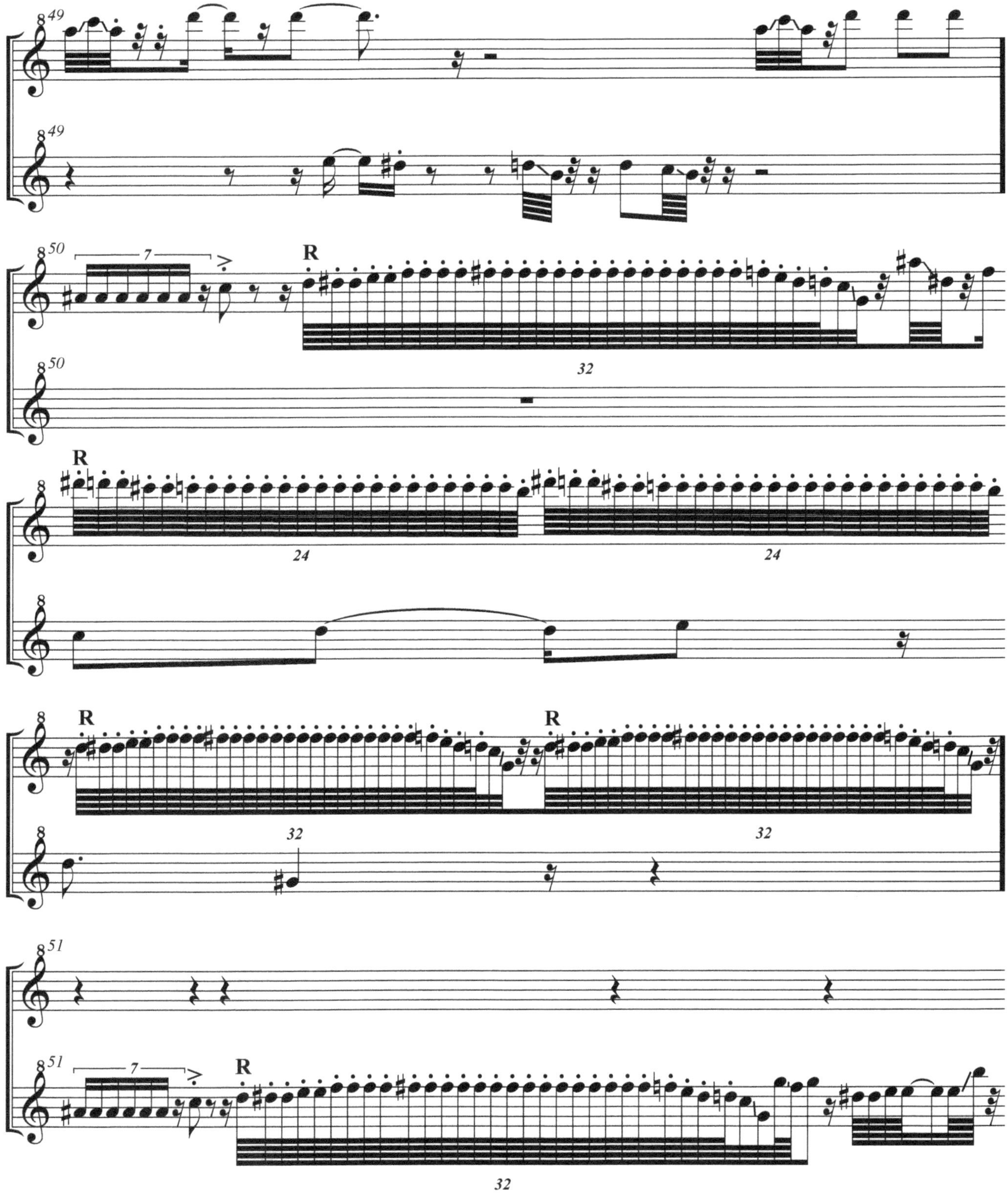

Example 8.10 *continued*

Example 8.10 *continued*

Acknowledgments

I am grateful to three wildlife recordists for their archival pied butcherbird recordings: composer David Lumsdaine for Pebbly Beach and Kinchega, ornithologist and forester H. Sydney Curtis for Witches Falls National Park, and Dr. Jenny Beasley for Green Lake. I also wish to thank two anonymous peer reviewers and Allison Chaplin, acquisitions editor at IUP. Finally, I bow to the pied butcherbirds featured in this volume (and the many others who are not but fully deserve to be). My understanding of their songs is partial, but my awe is absolute.

Appendix 1

*Audio Tracks**

AUDIO 1	Trephina Ridges Music Camp 0.0: Mount Benstead Creek, 2016
AUDIO 2	Trephina Ridges Music Camp 1.0, 2016
AUDIO 3	Trephina Ridges Music Camp 3.7, 2016
AUDIO 4	Trephina Ridges Music Camp 7.5, 2016
AUDIO 5	McDonald Street @ Bridge Street, Charters Towers, 2014
AUDIO 6	Chloride Street @ Millett Street, Charters Towers, 2014
AUDIO 7	McPherson Street, Charters Towers, 2014
AUDIO 8	Rutherford Lane, Charters Towers, 2014
AUDIO 9	Araluen Arts Centre, Alice Springs, 2014
AUDIO 10	Araluen Arts Centre, Alice Springs, 2017
AUDIO 11	Newhaven Wildlife Sanctuary, 2010
AUDIO 12	Erija Street, Alice Springs, 2016
AUDIO 13	Erija Street, Alice Springs, 2017
AUDIO 14	Greens Park, Georgetown, 2014
AUDIO 15	Greens Park, Georgetown, 2015
AUDIO 16	Cameron Road, North Queensland, 2010
AUDIO 17	Cameron Road, North Queensland, 2013
AUDIO 18	Fegen Road, Brisbane, 2010
AUDIO 19	Goldfields Caravan Park, Georgetown, 2013
AUDIO 20	Midway Van Park, Georgetown, 2013
AUDIO 21	Normanton Street, Georgetown, 2015

* Audio files are available at https://www.piedbutcherbird.com.

AUDIO 22 Archery Track, Alice Springs, 2010
AUDIO 23 Arltunga Turnoff, Ross Highway, Central Australia, 2022
AUDIO 24 Bullen Road 2.5, Alice Springs, 2020
AUDIO 25 Oonchiunpa Road, Alice Springs, 2021
AUDIO 26 Bitter Springs Creek, Central Australia, 2014
AUDIO 27 Palm Valley, Central Australia, 2014
AUDIO 28 Kinchega, New South Wales, 1984
AUDIO 29 Tibooboora, New South Wales, 2012
AUDIO 30 Stockyard Creek, North Queensland, 2010
AUDIO 31 Booth Road, Townsville, North Queensland, 2006

Appendix 2

Notation Conventions

My method proceeds by way of ear/eye/ear. I listen, I examine each phrase in the sonogram window, and then I import the birdsong audio file into my music notation program, which allows me to reconfirm each phrase as I proceed in matters of pitch, rhythm, and metronomic marking.

Most of my transcriptions employ a 4/4 time signature. The choice between simplifying for legibility and detailing the most exact rhythm possible is a vexing one; in the end, as stated earlier, complex rhythms are best represented via organic gesture by the human performer, even if they deviate slightly from the transcription. In addition, even the birds vary their rhythms somewhat, as documented in these transcriptions.

Rest signs mark out interphrase intervals. Some measures are split across lines due to length, with no bar line at the end of the first line. Phrases typically have variations, which I am keen to distinguish. However, I consider very slight variations of the same phrase as repetitions, since a microphone or bird's (often changing) position, as well as environmental conditions and diverse recording techniques from multiple recordists and a range of equipment, all introduce more variation into recordings than exists among the phrases as delivered. When an assumed note from another species turns up in the next iteration of that phrase, and the next, a revision is required—it was not another species after all. The process owes everything to a willingness to tack back and forth.

For the most part, I avoid indicating dynamics, which are subjective assessments under the best conditions. Articulations are similarly subjective, although I do indicate the most robust examples. For legibility, I notate my transcriptions an octave below the pitch where they were delivered, setting them in a treble clef with an "8" affixed above it.

Although rattles (a rapid succession of short notes whose timbre ranges from broad spectrum and harsh to hollow sounding), trills, and other quickly iterated decorations are audible at their original speed, I can measure the exact number of iterations in the sonogram window if I want that level of description. Otherwise, a tremolo expression mark can be applied. The fastest repeated iterations are marked as rattles (R for a rattle and QR for a quasi-rattle whose iterations do not fully separate); other slower and shorter repeated pitches could be marked as rattles, although I generally do not.

For accidentals, I prefer sharp signs but do draw on flats when a broad-spectrum note encompasses several pitches. For microtonal phrases, I place half-sharp accidentals above the notehead, making an already sharp note three-quarters sharp with the addition of a half-sharp sign, and even within the measure, the half-sharp sign applies only to that note unless tied.

I follow octave leaps and other intervals that interest me (like two ascending perfect fourths) as well as novel rhythmic patterns.[1] I tally motifs and variants that are held in common among individuals. Pacing; ostinatos; transposition; mnemonic sounds like *chip*, *wow*, *tik*, *tok*, and others; and chimeric sounds imported from another species and pasted into pied butcherbird song (see Irvine Street in chap. 7)—all of these get documented. Many of these aspects I speak into the recorder at the end of the song while my mind is fresh.

Since birds sing from their syrinx, which consists of a valve in each of two air tubes (bronchi) located just below the junction with the trachea, the singing of double notes is possible, including broad-spectrum or "noisy" notes. To notate a "noisy" note that lacks a clear pitch focus, I may replace a notehead with an X on the closest apparent pitch (or pitches) while I represent broad-spectrum notes that are less noisy with adjacent semitone noteheads.

To notate a portamento, I draw a line between the beginning and end notes. If to my ear either end dominates, I use a normal notehead to indicate that end, while a grace note sometimes serves for the other end of the portamento. The speed is usually swift and even, and I do not attempt to measure that aspect. Whether the steepness or the speed of a portamento is more significant than the pitch to these birds is unknown.

Notation is not the full story. In a supplemental summary sheet, I follow parameters that are not, or not easily, notated. I may tally the number of rattle types and rattle units, the rattle pulse rate, and average rattle frequency; the lowest and highest frequencies in the song; whether phrases fit twelve-tone equal-tempered reasonably well or not; whether there is a sense of a tonic or a major or minor key; whether notes are pure whistles or more broad spectrum sounds; the widest ascending and descending leaps; and the number of portamento types.

My transcriptions cleave to a two-sided coin: On the one side, I make every attempt to detail avian achievement and nuance, while on the other, I simplify to make the notation accessible and legible to human musicians.

Abbreviations

km	kilometer
QR	quasi-rattle
R	rattle

Note

1. For more on the rhythm and pacing of pied butcherbirds, see Xing et al., "Syntactic Modulation of Rhythm."

Bibliography

Adelaide Festival. "KRONOS Five Decades." Accessed July 7, 2024. https://2023.adelaide festival.com.au/events/kronos-five-decades.

Andersen, Johannes C. *Bird-Song and New Zealand Song Birds*. Whitcombe & Tombs, 1926.

"Australian Wildlife Sound Recording Group." Accessed July 8, 2024. https://awsrg.org.au.

Bateson, Gregory. *Steps to an Ecology of Mind: Collected Essays in Anthropology, Psychiatry, Evolution, and Epistemology*. Jason Aronson, 1972/1987.

Curtis, H. Sydney, and Hollis Taylor. "Olivier Messiaen and the Albert's Lyrebird: From Tamborine Mountain to Éclairs sur L'au-Delà." In *Olivier Messiaen: The Centenary Papers*, edited by Judith Crispin, 52–79. Cambridge Scholars, 2010.

Hindley, David. "The Music of Birdsong." *Wildlife Sound* 6, no. 4 (1990): 25–33.

Janney, Eathan, Hollis Taylor, Constance Scharff, David Rothenberg, Lucas C. Parra, and Ofer Tchernichovski. "Temporal Regularity Increases with Repertoire Complexity in the Australian Pied Butcherbird's Song." *Royal Society Open Science* 3, no. 160357 (2016): 1–15.

Kircher, Athanasius. *Musurgia Universalis Sive Ars Magna Consoni Et Dissoni*. Franceso Corbelletti, c. 1650.

Knobloch, Eberhard. "*Musurgia Universalis:* Unknown Combinatorial Studies in the Age of Baroque Absolutism." *History of Science* 17 (1979): 258–75.

Mâche, François-Bernard. *Music, Myth and Nature*. Translated by Susan Delaney. Harwood Academic, 1983/1992.

Martinelli, Dario. *How Musical Is a Whale?: Towards a Theory of Zoömusicology*. International Semiotics Institute, 2002.

Jon Rose. "Night Songs." Accessed July 8, 2024. https://www.jonroseweb.com/f_projects_night_songs.php.

Queensland Government. "Banana, Queensland." Accessed July 14, 2024. https://www.queensland.com/au/en/places-to-see/destination-information/p-596c087d1886d3db60a48ef3-banana.

Ratner, Leonard G. "*Ars Combinatoria*. Chance and Choice in Eighteenth-Century Music." In *Studies in Eighteenth-Century Music: A Tribute to Karl Geiringer on His Seventieth Birthday*, edited by H. C. Robbins Landon and Roger E. Chapman. Oxford University Press, 1970.

Seashore, Carl E. *Psychology of Music*. Dover, 1938/1967.

Sethares, William A. *Tuning, Timbre, Spectrum, Scale*. Springer, 1999.

Sotavalta, Olavi. "Analysis of the Song Patterns of Two Sprosser Nightingales, *Luscinia Luscinia*." *Annals of the Finnish Zoological Society "Vanamo"* 17, no. 4 (1956): 1–31.

Sundberg, Johan. "The Perception of Singing." In *The Psychology of Music*, edited by Diana Deutsch, 171–214. Academic Press, 1999.

Szőke, P., W. W. H. Gunn, and M. Filip. "The Musical Microcosm of the Hermit Thrush: From Athanasius Kircher's Naive Experiments of Musical Transcription of Bird Voice to Sound Microscopy and the Scientific Musical Representation of Bird Song." *Studia Musicologica Academiae Scientiarum Hungaricae* 11, no. 1/4 (1969): 423–38.

Tate, Henry. *Australian Musical Possibilities*. Edward Vidler, 1924.

Taylor, Hollis. "Birdsong: Language or Music?" *Unlikely* 8 (2023): n.p. https://unlikely.net.au/issue-08/birdsong.

Taylor, Hollis. "Evidence for Teaching in an Australian Songbird." *Frontiers in Psychology* 12, no. 449 (February 22, 2021): 1–22. https://doi.org/10.3389/fpsyg.2021.593532

Taylor, Hollis. "How Musical Are Animals? Taking Stock of Zoömusicology's Prospects." *Music Review Annual* 1 (2020): 1–35.

Taylor, Hollis. *Is Birdsong Music? Outback Encounters with an Australian Songbird*. Indiana University Press, 2017.

Taylor, Hollis. "Messiaen's Australian Birds: The Impact of H. Sydney Curtis." *Musicology Australia* 44, no. 1 (2022): 58–78. https://doi.org/10.1080/08145857.2022.2111594

Taylor, Hollis. "Olivier Messiaen's Transcription of the Albert's Lyrebird." *AudioWings* 12, no. 1 (June 2009): 2–5.

Taylor, Hollis. "Whose Bird Is It? Messiaen's Transcriptions of Australian Songbirds." *Twentieth-Century Music* 11, Special Issue no. 01 (2014): 63–100. https://doi.org/10.1017/S1478572213000194

Tiessen, Heinz. *Musik Der Natur*. Atlantis, 1953.

Xing, Jeffrey, Tim Sainburg, Hollis Taylor, and Timothy Q. Gentner. "Syntactic Modulation of Rhythm in Australian Pied Butcherbird Song." *Royal Society Open Science* 9, no. 9 (2022): 220704. https://doi.org/10.1098/rsos.220704

Index

Adelaide Festival, 71
Andersen, Johannes C., 5
antiphon/antiphonal song, 204, 245, 246. *See also* duet/duo; ensemble song
arpeggio, 171
ars combinatoria, 168
Australian Wildlife Sound Recording Group, 171

Bateson, Gregory, 4
birdsong, 1, 7, 70, 72, 73, 121, 122, 168–69, 171, 203–4; female song, 245; notation of, 1–5, 171, 289–90; vocal repertoire, 6. *See also* antiphon/antiphonal song; duet/duo; ensemble song; mimicry

calls, bird, 1, 2, 73, 171, 203, 247
chromatic, 1, 246
combinatoriality/combinatorics, 1, 7, 73, 122, 168, 171, 246
countersong/countersinging, 246
crescendo, 203
Curtis, H. Sydney, 246, 285

duet/duo, 3, 122, 245–284. *See also* antiphon/antiphonal song; ensemble song

ensemble song, 1, 3, 72. *See also* antiphon/antiphonal song; duet/duo

fanfare, 4, 73, 204

half-sharp, 4, 290
Hindley, David, 5
hocket, 245, 246

intonation, 4

Kircher, Athanasius, 5, 168

Mâche, François-Bernard, 5n4
major/minor, 121, 171, 290
Martinelli, Dario, 5n4
melisma, 246
melodic contour, 4, 204
Messiaen, Olivier, 5
meter, 4. *See also* rhythm
microtonal/microtone, 4, 290
mimicry, 73, 171
mnemonic sounds, 290

octave, 73, 204, 246, 289, 290
ostinato, 37, 290

pied butcherbird: 1–5, *3*, 6–8, 37, 70, 72, *169*, *170*, 203–4, 245–47; lifespan of, 70; repertoire of, 6; speciation time, 168; subspecies *Cracticus nigrogularis*, 1, 168–202; subspecies *Cracticus nigrogularis picatus*, 168, 203–44; vocal range of, 122; web link to, 4, 6
portamento, 4, 5, 171, 203, 290

rattle/quasi-rattle, 73, 171, 246, 289, 290
rhythm, 4, 121, 168, 289, 290
Rose, Jon, 71, 204

songpost, 6, 7, 71, 73, 122, 171, 245, 246
sonogram, 5, 289
Sotavalta, Olavi, 5
Susskind, Walter, 122
syrinx, 290
Szeryng, Henryk, 122
Szöke, Péter, 5

Tate, Henry, 1
Tiessen, Heinz, 5
timbre/timbral, 1, 4, 37, 73, 122, 171, 289
time signature, 4, 289
tonic/tonal center, 38
transcription, 1–5, 7, 8, 121, 171, 203, 204, 246, 289–90
trill, 289
typology, 121

vocal learning, 2, 7

web link, 4, 6

zoömusicology, 2, 5n4

HOLLIS TAYLOR is a research affiliate at the Sydney Conservatorium of Music. She is a violinist, composer, zoömusicologist, and author of *Is Birdsong Music?: Outback Encounters with an Australian Songbird.*

For Indiana University Press

Sabrina Black *Editorial Assistant*
Tony Brewer *Artist and Book Designer*
Allison Chaplin *Acquisitions Editor*
Anna Garnai *Production Coordinator*
Sophia Hebert *Assistant Acquisitions Editor*
Samantha Heffner *Marketing and Publicity Manager*
Katie Huggins *Production Manager*
Gigi Lamm *Director of Sales and Marketing*
Alyssa Nicole Lucas *Marketing and Publicity Manager*
Darja Malcolm-Clarke *Project Manager/Editor*
Annie L. Martin *Editorial Director*
Dan Pyle *Online Publishing Manager*
Michael Regoli *Director of Publishing Operations*
Pamela Rude *Senior Artist and Book Designer*